Poke-yoke

Poka-Yoke

An Introduction to Poke-Yoke for Engineers

Copyright © 2016, First Edition Author: E.P. Tobin

All rights reserved.

CONTENTS

What is Poke-yoke?	1
A Brief History	3
Key Principles	4
A Roadmap to success	6
Customer Driven Companies	10
Are Errors avoidable?	13
Sampling plans	14
Sources of Errors	20
Jigs & Fixtures	23
Common Poke-yoke tools	24
Graphical Examples of Poke-yoke	29
Approaches to Mistake Proofing	46
Cost Savings with Poke-yoke	48
Right first time (RFT)	49
Plan-Do-Check-Act	51
5s	53

What is Poka-yoke?

Poka-yoke, is a technique for avoiding simple human error in the workplace. Simply put, it aims eliminate mistakes and is often referred to as mistake-proofing or fail-safe work methods.

Poka-yoke is simply a system designed to prevent inadvertent errors made by workers performing a process.

The word "Poka-yoke" is Japanese for mistake-proofing or mistake avoidance. It involves the design of products, work practices, fixtures and jigs etc. that either prevent the mistakes or errors that result in defects. A secondary aim of Poke Yoke is to make any defect easy to recognize with minimum time, skill and expertise.

It is accepted as a simple and inexpensive way of preventing defects from being made or identifying a defect so that it is not passed to the next operation, downstream process and ultimately, the consumer.

The benefits of Poke Yoke are very wide. The specific benefits depend on the nature of the work and also where the focus is placed when executing a Poke-Yoke program. If it focusses on cost reduction, the metric might be a decrease in set up times or processing times. If the focus is on quality, a redesign of jigs or fixture may be required. Here are 12 benefits of Poka-yoke:

- Reduce set-up issues
- Improve product quality
- Improve yield
- Reduce rework
- reduce manufacturing cost
- decrease set-up time
- decrease set-up complexity
- improve housekeeping
- remove dependence on high skill levels or experience
- increased manufacturing flexibility
- improved work attitudes
- reduce manufacturing cost

A Brief History of Poka-yoke

Poka-yoke was developed by Shigeo Shingo. It was he who wrote the defining works on this technique, although he is not the one who invented the idea. The concept of mistake proofing had been around for a long time before it was identified as a specific lean tool.

Initially many people called the technique fool-proofing, however this could at times be perceived as being derogatory towards the people using the device.

Key Principles of Poke Yoke

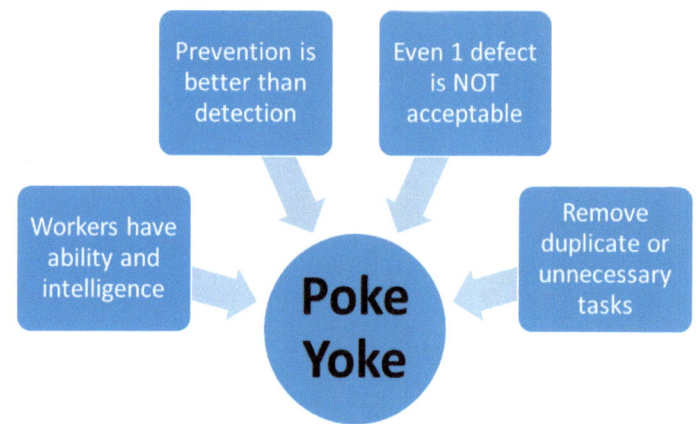

Workers have ability and intelligence...

For the application of any methodologies such as Poke Yoke, the right approach and attitude largely determines the success. The persons responsible for the rollout of such methodologies must fully support the program and truly believe in the benefits. Typically, Top Management are responsible for the rollout of programs such as Poke-Yoke and Lean etc. Human intelligence can quickly see through any new initiatives that are not understood or supported by management.

There should also be an acknowledgement of worker's ability and aptitude. No matter what the role of the person, be it factory operator, technician, electrician or engineer, trust and confidence must be given to all stakeholders. It is a dangerous trait to overlook the input and contribution of junior staff.

Prevention is better than detection...

Preventing defects saves money. Detecting defects costs money. Therefore, the preference is to prevent defects before they happen. This is where mistakes can be eliminated by Poke-Yoke techniques. If you eliminate mistakes or errors, this in turn works to eliminate defects. While inspection and detection systems will always have some use, the absence of defects will reduce the dependency of detection.

Even 1 defect is NOT acceptable...

Having Zero defects is the golden rule. If defects are accepted as part of a culture, then complacency follows.

Remove duplicate or unnecessary tasks...

It may seem like a trivial observation, however, allowing duplicate tasks that are unnecessary (not required) can lead to mistakes causing defects. Why you may ask? Apart from the cost of completing extra manufacturing steps that are not required, extra steps or tasks generally result in extra handling. Extra handling may involve transfer to and from work stations, labelling, paperwork completion and stand down times. All of these activities can be a source of error generation

A Roadmap to Success

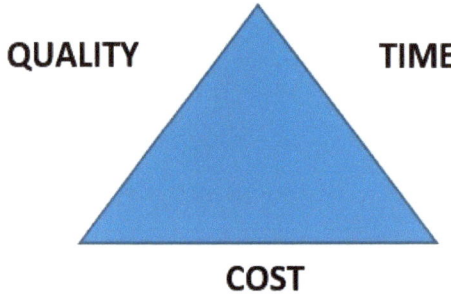

Figure: Adapted representation of the Project Management Triple Constraint.

The above figure shows the 3 elements of the triple constraint theory often used in Project Management. This theory is also applicable to manufacturing and mistake proofing.

Cost

Cost is one of the most important factors to today's consumer. Especially when faced with a choice of manufacturers, the sale price needs to be in keeping with other manufacturers. Even if the product is of superior quality and function, cost needs to be understood and managed effectively.

With regards to Poke Yoke and mistake proofing, there is an opportunity to reduce costs during manufacturing. If errors and mistakes are prevented, this reduces defects. Not having to rework product of dump defective products helps to save money and frees up cash within a factory.

If you can reduce manufacturing costs by eliminating defects and wastage, this creates an opportunity to make a higher profit margin on the market o retail price.

Quality

Some defects may result in dissatisfaction to the customer. More serious defects may prevent the customer using the product at all or in a safe manner. It is in everyone's interest to manufacture a quality product that meets the user's requirements. Product Quality is often a byword for safety which should always be priority to a manufacturer.

It should be understood that changes in the level of product quality may impact upon both costs and time. It can be said that this is a balancing act, however, there should be a minimum standard of quality that is met every time.

Time

"Making mistakes increase the amount of work."

The time constraint also has implications of cost and quality. Take a manual process such as hand finishing wood. If the job is rushed or time is squeezed, the quality of the finish may not be up to standard.

If the hand finishing takes too long, it impacts upon the overhead costs associated with labour and the cost to deliver the product rises. It may be in the interested to reduce the time required to complete a task or job, but the quality of work must be maintained.

The time to complete each manufacturing step all contribute to the final cost of the product. This creates a lead time for each product type. Lead times become important in scheduling the right products and right volumes in order to meet the market requirements.

In recent years, with many consumer products, people expect next day delivery. This means the manufacturer most also be responsive and flexible to demands of the market and wholesalers.

Customer Driven Companies

Every manufacturer wants to meet the expectations its customers in Quality and other factors such as delivery times and costs. A popular standard which is used by manufactured worldwide is ISO 9000 Quality Management.

The ISO 9000 is made up of the following standards:

- ISO 9001:2015 - sets out the requirements of a quality management system
- ISO 9000:2015 - covers the basic concepts and language
- ISO 9004:2009 - focuses on how to make a quality management system more efficient and effective
- ISO 19011:2011 - sets out guidance on internal and external audits of quality management systems.
- (Ref: http://www.iso.org/iso/iso_9000)

A key requirement of ISO 9000 is Customer focus, the requirements of clause 5.2 deals with meeting customer requirements, and also managing the feedback from customers on an on-going basis. The corner stone of

Requirements of "continual improvement" and "customer satisfaction"

Customer driven Practices and Quality Policies

A quality policy is a concise statement that sets out a company's commitment to the customer and the commitment to delivering quality products and services. Often a Quality Policy will be displayed in the reception area of a company or is available to download as a document on their website.

The quality policy must be relevant to the business operations. While many themes and traits are common across different sectors, the quality policy for a service company would likely differ slightly to a company that manufacturers physical products.

Example of a Quality Policy

"We practice continual Improvement to achieve customer delight by providing Customer-Centric, Cost-effective, Timely and Qualitative software solutions. We are committed to meeting the regulatory requirements of Medical Device manufacturing, and meeting our customer expectations"

Are Errors avoidable?

Can errors be eliminated?

In the journey to achieve zero mistakes and zero defects, one school of thought believe mistakes that people make can be reduced to a minimum or indeed eliminated. There are several factors that help reduce the amount of mistakes people make. Training and experience are key parts, along with the proper systems and resources e.g. tooling, instruction, work area setup to name a few.

Are errors inevitable?

An opposing view of errors is that people always make mistakes, no matter how small or low in occurrence. Even if we accept mistakes as a part of life, we still tend to blame the people who make them. The risk of adopting this philosophy is that defects can be missed during manufacturing which can result in defective products been sold commercially.

Sampling Plans

Dealing with Errors

Remember, the preferred method when it comes to errors is not to make them by mistake proofing manufacturing processes.

The second method of preventing errors is by inspection.

Why sampling plans?

Financial costs are always associated with inspection regimes. Take traditional manual inspection. This requires a factory operator to be available to inspect products. Furthermore, time and resources are required to train personnel in the inspection procedure. Add to this the time required to complete paperwork and the cost of inspection can be substantial. Even if technology can be used to automate or help the inspection process. This can reduce inspection costs, increase detection but often require large capital costs and ongoing technical support.

Acceptable Quality Levels (AQLs)

Some plant managers say, "It would take us all day to inspect all our products. There may be a few defects, but sampling is still the most practical way to check.

An Acceptable quality level (AQL) of 0.1 percent is applied to all our products. AQL sampling inspection provides a level of protection for the manufacturer but as much from the consumer's perspective. A 0.1% defective rate will result in one customer receiving a defective product.

American National Standards Institute (ANSI)

Sampling Procedures and tables for Inspection are commonly taken from the American National Standards Institute (ANSI)/American Society for Quality (ASQ). The standards provide defined tables detailing a range of inspection regimes. It also provides some simple instructions on how to correctly select the sampling plan based on the population size and the acceptable risk.

Statistical Sampling

Feedback works to reduce the cost of inspection, as issues are reported upstream and fixed. However, the speed at which feedback is sent and acted upon determines if feedback can be an effective. If the time between the feedback and fix is large, defects will still be proceeded.

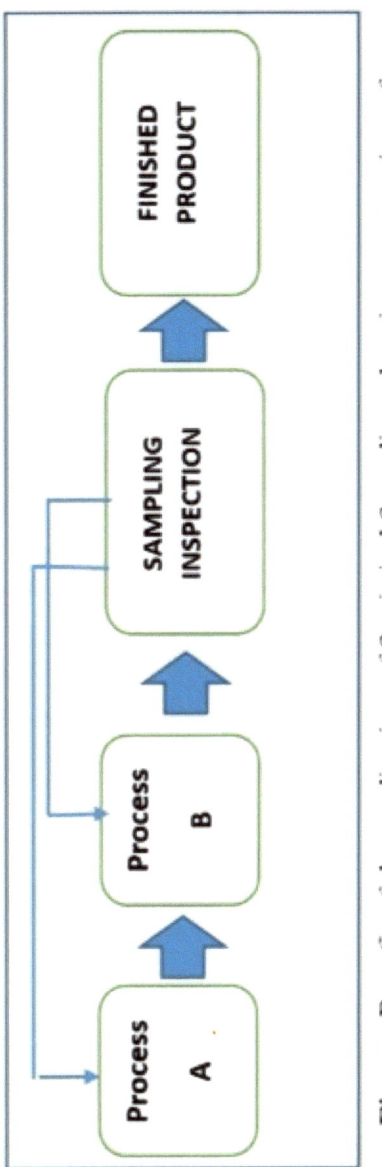

Figure: Benefit of the application of Statistical Sampling, showing an overview of a manufacturing process with sampling inspection prior to final assembly.

100 % inspection (Traditional Inspection)

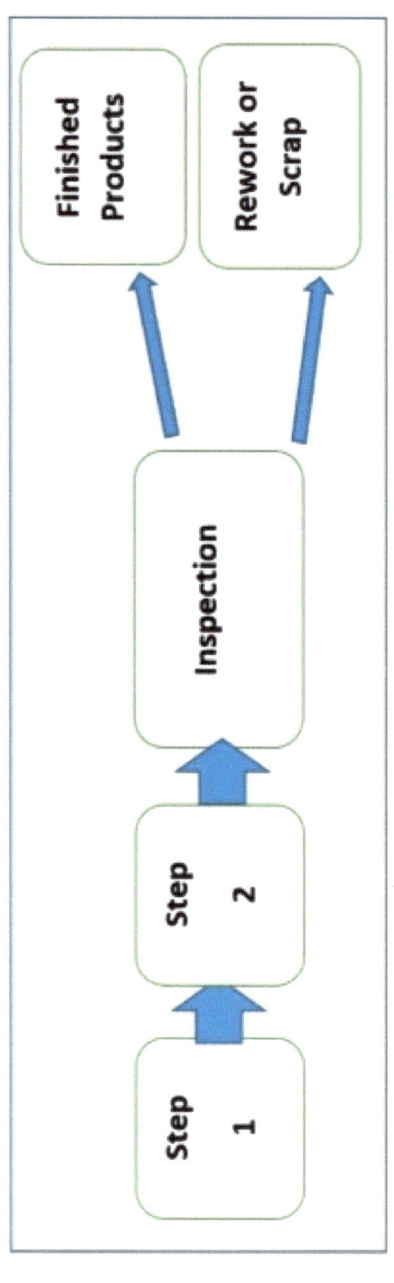

100 % inspection (Traditional Inspection)

100% inspection as the term suggests involves the inspection of each and every part. If it is a manual process, it is often time consuming and therefore costly. The risk with 100% inspection is that bad product may not be detected and makes its way to the customer. In contrast to in process inspection sampling is that it does not prevent defects from been manufactured. This can lead to a build-up of defective products which increase costs.

200% Inspection is often a term used to describe a process where the product is inspected twice, perhaps by different people or after a hold time. This risk with repeated inspection processes it that at each inspector depends on the other. Which may result in two ineffective inspection stages.

Sources of Errors

Poke-yoke tries to prevent or eliminate human errors. There are several common types of Human errors some of which include:

Communication errors: Often mistakes occur due to a lack of communication or due to someone misinterpreting instructions. *Mitigation:* Insure any critical communications are written or available to review if there is doubt.

Rookie Errors: Sometimes we make mistakes through lack of experience. For example, a new worker does not know the operation or is just barely familiar with it. *Mitigation:* Skill building and work standardisation.

Compliant Errors: If no consequences are perceived sometime we can overlook steps or processes. *Mitigation:* Foster a sense of personal responsibility and the impact of small defects on the customer.

Forgetting: Humans are prone to forgetting steps or tasks especially if they are repetitive and they are working on the same processes for long periods of time. *Mitigation:* Provide Checklists to operators and workers in order to formalise the process. Paperwork documenting critical steps will alert the operator if they forget a step.

Poke-yoke

Procedure related errors: If instructions or standard operating procedures are inadequate it may lead to errors.

Mitigation: Ensure existing work instructions are accurate and reflect the proper and necessary actions for safety, quality and prevention of mistakes.

Sources of Defects

There are various types of defects. The table below lists some common defects along with some suggested sources:

1.	**Omitted processing steps:** **Time pressures** **Carelessness** **Lack of training**	6. **Processing wrong workpiece:** **Workflow layout errors**
2.	**Processing errors:** **Skill deficiency** **Wrong specification**	7. Misoperation: Poor equipment / tooling Operator error
3.	**Errors setting up work pieces:** **Lack of training / experience** **Lack of correct tools**	8. Processing errors: Operator error Wrong machine
4.	**Missing parts:** **Oversight** **Workflow issues**	9. Equipment setup errors: Lack of training Wrong settings
5.	**Wrong parts:** **Parts not identified**	10. Tools and jigs: Grease of coolant not applied

	Wrong jigs or tooling

Jigs & Fixtures

The terms "Jig" and "fixture" are commonly used in the manufacturing industry particular in CNC machining and fabrication. Many machining processes require jigs and fixtures in order to achieve consistent and accurate results.

Jig

A jig is used to guide the item or component that has to be machined while a fixture holds in place of "fixes" the component to be machined or processed.

Fixture

A fixture is used to hold the component or part during the machining process. Its purpose is ***not*** to guide the part towards the machining tool. Fixtures are secured with the table surface of the mills in most of the cases. Fixtures reduce the need for other tools and facilitates more accurate machining and processes.

Common Poka-Yoke tools

Just to recap quickly, by using Poke-yoke tools, we are trying to eliminate human errors. These errors are usually done mistakenly due to poor judgement or concentration.

Poka-yoke helps prevent defects resulting from human error or mistakes. Human factors or human errors can lead to quality defects. Poke Yoke helps people including factory operators, fabricators, assembly personal and engineers to reduce defects due to errors. Some common examples ok Poke Yoke tools include:

1. Checklists

Checklists are a practical and efficient way of detecting errors before they impact a process or product. Some of the best Checklists are designed to be completed in a relatively short period of time, however, this can be influenced by the complexity of the task at hand. The principle still stands that designing a checklist that is to the point and easily completed will deliver the best benefits. Checklists should focus on factors that if overlooked in error can lead to defects.

2. Error detection - Visual and Audio alarms

A lot of automated equipment has in-built controls that will alarm visually and audibly when a process begins to drift of go out of control. For example, a parts washer may have temperature alarms to indicate if the temperature drops or rises below the process settings. Alarms and warning systems therefore can prevent mistakes and defects before they materialise. They can also help detect when errors do occur and help to ensure the customer gets a quality product.

3. Guide pins

Guide pins are a proven way to help force the proper set-up and assembly of parts. Typically guide pins of different sizes are used to orient and position components in the desired manner. This prevents misoperations such as drilling or machining in the wrong position. Guide pins also help to ensure components are assembled in the correct way.

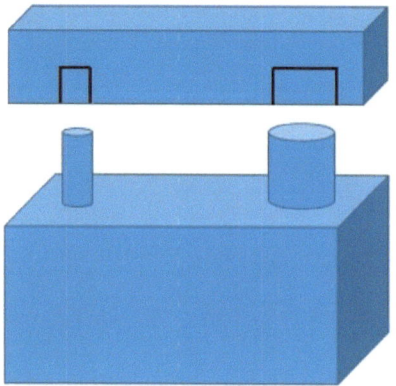

4. Jigs

As previously described, a jig is used to guide the item or component that has to be machined while a fixture holds in place of "fixes" the component to be machined or processed. Jigs are very common and if quality upfront engineering to given to jig design, it can eliminate a lot of errors and defects during manufacturing, especially at high volumes.

Graphical Examples of Poke-yoke in Practice

Graphical Examples

The best way to understand the application of Poke-yoke is to look at some examples of everyday items that have been designed with mistake proofing in mind.

Imagine a scenario where you receive a new desktop computer. Many of the external devices associated with desktop PCs still use cables to connect to the desktop. Taking a simple setup consisting of a keyboard, mouse, monitor and power supply. This would indicate four cables that require connection. We generally do not think that this type of setup is difficult or technical, as the cables connections are different shapes and typically colour coding is also used. However, if all the cable connections were the same shape and size, the task of connecting all the external devices to the desktop would be not as easy.

Below on the Left hand side is a simple representation of a VGA cable traditionally used to connect a visual display unit to a device or PC. Using a unique shape or profile for the connection forces the user to connect it in one way only. If the port was simply rectangular, it is likely to cause confusion and damage when connecting.

Figure: Simple examples of Poke Yoke

A similar example shown above in the middle is an audio cable. The diameter of the inner pin (shown in yellow) and the diameter of the recess ensures that the cable can only be connected to the matching port.

Finally, on the right hand side is the familiar shape of a sim card. The chamfer at the top left hand corner of the sim card ensure the user orientates the card in the correct manner when inserting it into a mobile phone.

Poke-yoke

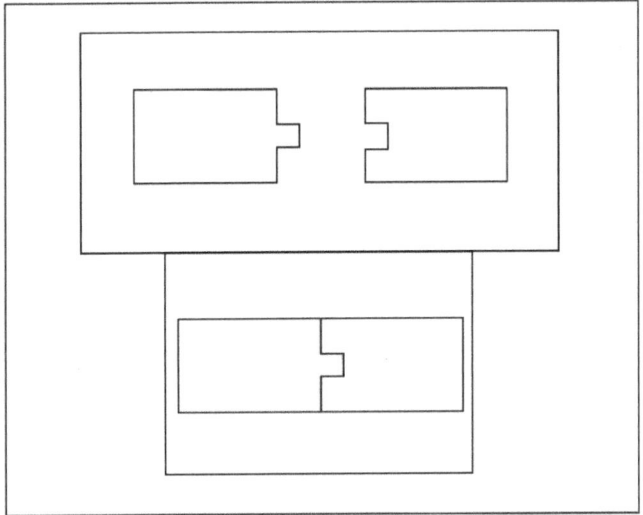

Figure: A simple design that forces the correct assembly of two parts.

Worked Examples

Example 1: Poke-yoke device to ensure a minimum fill level

Overview: The above container requires a minimum amount of tablets to be added during the packaging process. Automation is typically the best way to control the "fill amount". The dispensing mechanism could be controlled by a counter, a hopper or a magazine. Automation can also be used to check the weight of the filled container. Alternatively, a vision system might be used to ensure the bottle is filled to the right level.

However, a level of in-process inspection by an operator may be required also.

Before Improvement: Unaided, an inspector or operator may find it difficult to determine if each container meets the minimum fill level.

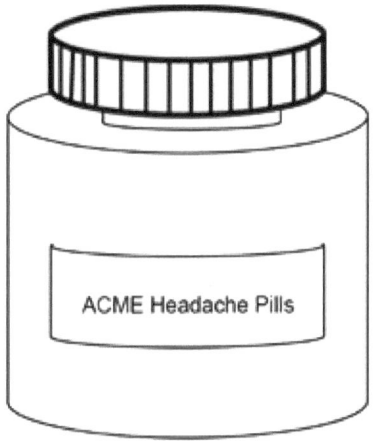

After improvement: Using an inspection aid (shown below) the inspector or operator can get a more accurate fill level. In this instance, the inspection jig must be design to highlight the minimum fill level on the container. Therefore, the depth at which the bottle sits into the jig pocket is critical.

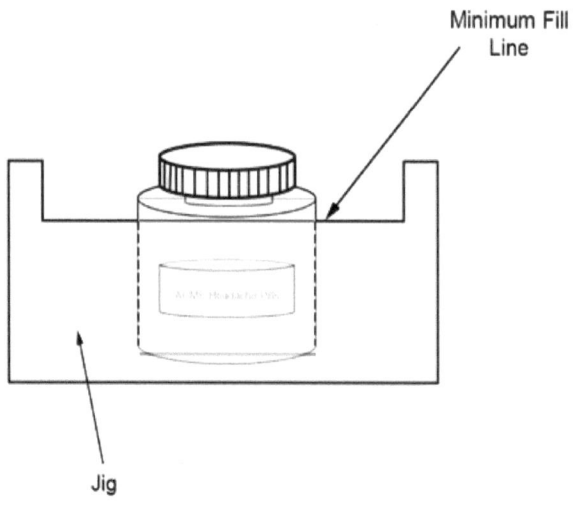

Figure: After implementation

Example 2: Contents to Box Verification

Overview: After a simple packaging process, the Carton should be packed with the product at the end of the line. The manufacturer wants to avoid shipping boxes that are not packed with product. There are two steps that the manufacturer would like to achieve. (1) Detection of empty cartons and (2) removal of empty cartons. As with modern technology, automation can provide complex and very effective ways to achieve the above results.

Before Improvement: The carton needs to be inspected manually by the operator in order to determine if the box contains the product. This would be a time consuming process and may damage "good" product also.

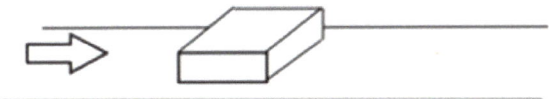

Figure: Before Improvement

After improvement: Using a simple air jet, any boxes that are empty will be ejected off the line into a rework or waste bin.

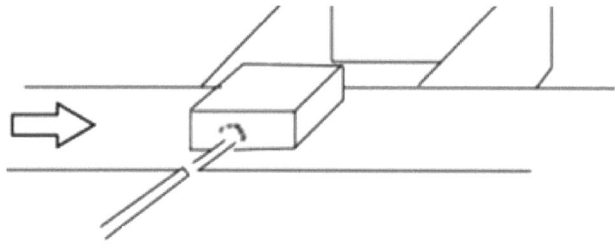

Figure: After improvement

Example 3: Machine Guarding

Overview: This example illustrates how Poke-yoke can be used to ensure the machine guarding is assembled correctly.

Before Improvement: The below metal component when is place and fixed to a machine acts as a safety guard to prevent operator access. The correct orientation of the guard when fixed in place. The four points of fixation are illustrated by the circles below. Each hole is the same diameter and the same size bolt is used for each position.

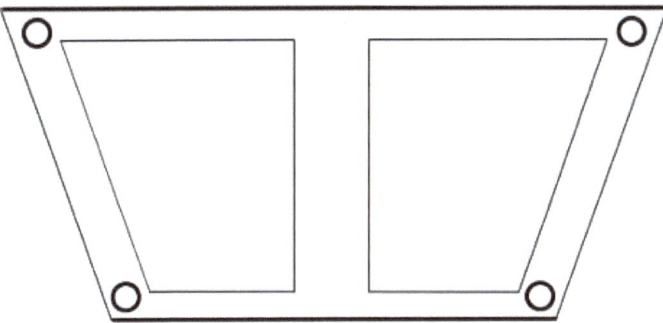

Figure: Before Improvement

After improvement: In order to ensure the guard is placed in the correct upright position, the diameter of the two bottom holes have been reduced to a smaller diameter. This means there is only one way that the guard can be fixed to the machine.

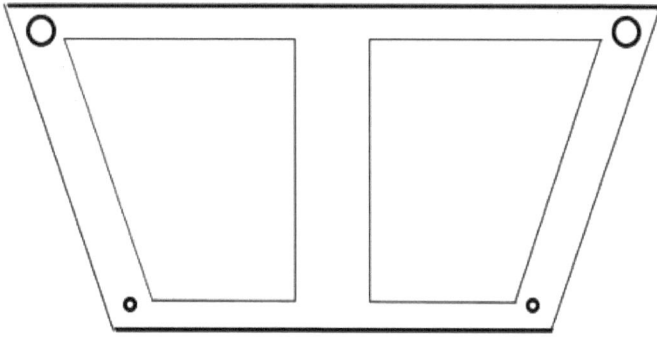

Figure: After improvement

A different but similar Poke-yoke is illustrated below where instead of modifying the hole diameter, the position of the holes at the opposing ends of the part are different. Again this design facilitates the correct orientation of the part to frame.

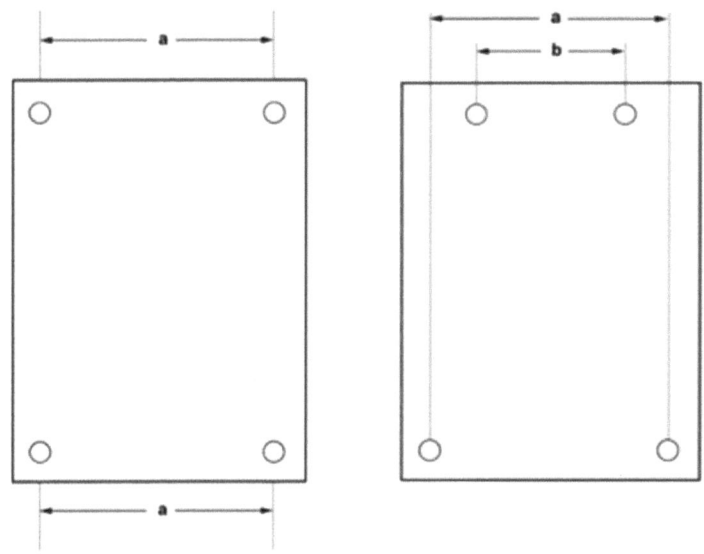

Figure: Right Hand Side drawing (Before), Left hand side (after)

Example 4: Base-plate drilling

Overview: During a machining job, a baseplate needs have 5 holes drilled in each one. Due to the high volume of base-plates been processed, sometimes only 4 of the five holes are drilled. This leads to unacceptable levels or rework.

Before Improvement: The drill operator is required to count manually the number of holes drilled per base-plate.

Figure: Before Improvement

After improvement: A new storage method for base-plates after they are drilled is introduced. The new storage method requires the operator to stack the drilled base-plates on a simple 4 column holder. If a hole is missing on a base-plate, the part cannot be stacked and cannot move on to the next machining step.

Poke-yoke

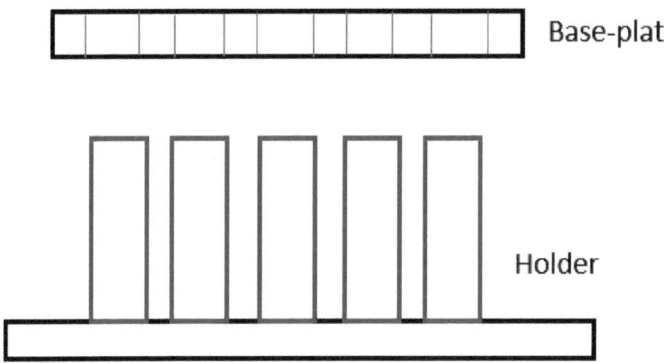

Figure: After improvement- Base-plates can be stacked on the holder if all holes are present.

Example 5: Drilling operation

Overview: In this example, the drilling operation must drill to a specified depth, and must not exit the base of the component.

Before Improvement: The operator may extend the drill piece too far vertically.

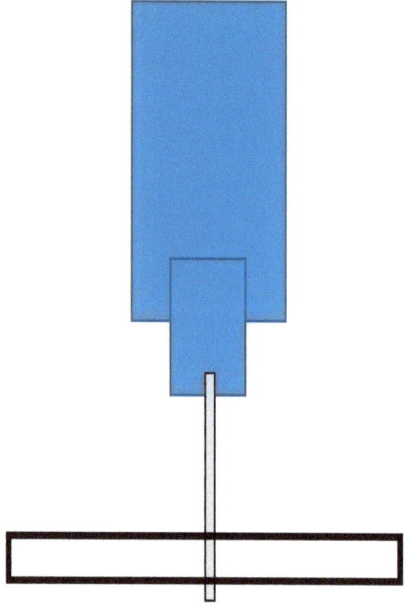

Figure: Before Improvement, showing drill bit exiting the component.

After improvement: A mechanical stop is fitted to the drill head (shown in grey). This will only permit the travel to the desired hole depth.

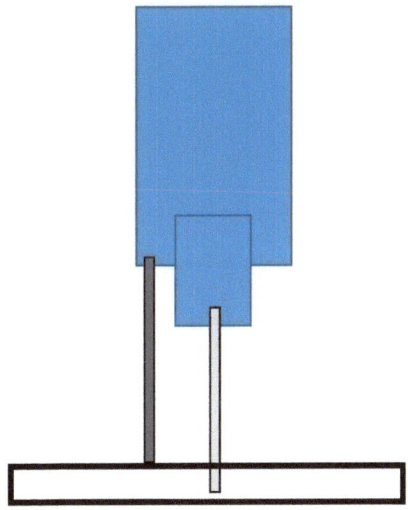

Figure: After improvement with mechanical stop fitted.

Example 6: Assembly of moulding die's

Overview: The correct assembly of the dies requirements the pocket to be on the right of the part. Due to the pins been the same size, there can be issues with incorrect assembly and setup.

Before Improvement:

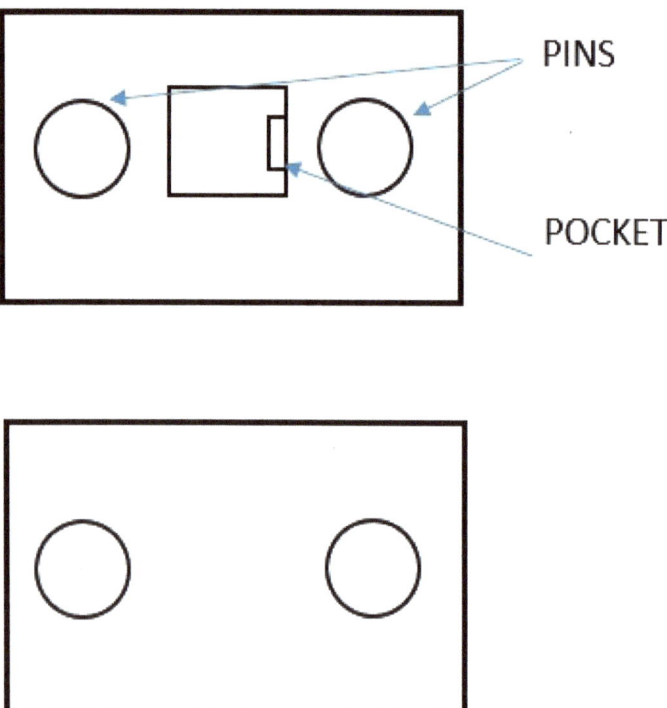

Figure: Before Improvement

After improvement: The pin diameter on the right hand side is smaller than the other pin.

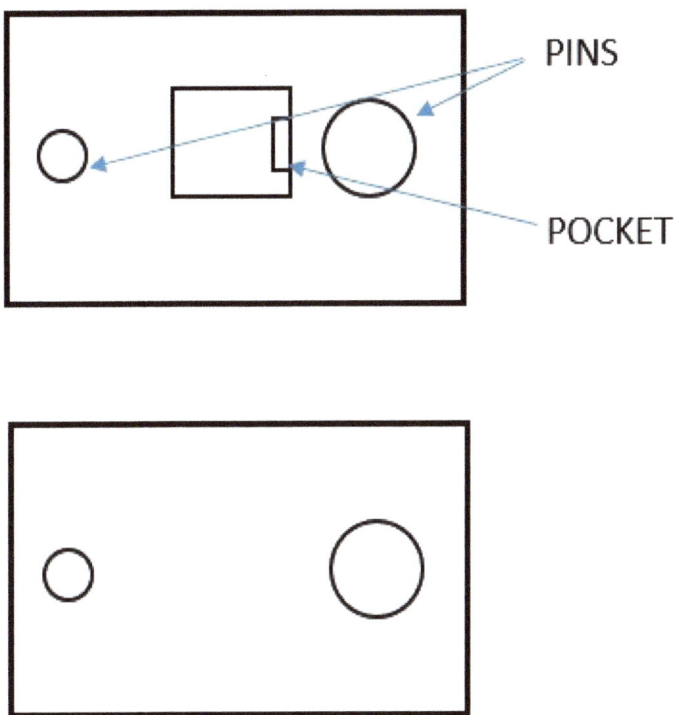

Figure: After improvement showing pins of different sizes.

Approaches to Mistake Proofing

As previously discussed, mistake proofing relies on prevention and detection. By far the most cost effective and preferred strategy is prevention. Prevention means that waste is avoided and the need and volume or rework is absent. Detection ensures that the process can be corrected and defective products do not reach the customer.

Method	Prevention	Detection
Shutdown	Before a mistake is about to happen	After a mistake has been made
Control	Errors cannot happen	Defective components are stopped
Warning	Before the process changes	When the process changes

5 Steps for Mistake Proofing

STEP 1: Describe the defect or potential defect. There may be more than one defect so it is best to separate this out as they may have different causes. If you are trying to prioritise the most important defect to prevent from occurring, there are four simple options (1) look at the defect rates or (2) look at the most critical defects that effect customer. (3) what defects are the most expensive (4) what defects take the longest to correct or rework.

STEP 2: Identify the process step where the defect is made or happens.

STEP 3: Observe the process steps from the start of the process to the point at which the defect is made. Compare the results to the process steps or operations.

STEP 4: Make a list of the potential errors. E.g. man, machine, method, setup.

STEP: 5 Identify the source of the defects

STEP: 6 Design a Poke-yoke device in order to prevent of detect the defects.

Cost Savings with Poke-yoke

It is logical to see that preventing defects before they occur is a powerful way to reduce costly rework and scrappage of products. The above illustration highlights the increasing costs associated with dealing with defects and mistakes as you move down stream in the process and towards the finished product and shipping.

A full cost-benefit analysis or risk-benefit analysis should be carried out when looking to implement a Poke Yoke device.

Right First Time (RFT)

Right First Time strives to create a culture of excellence. People are challenged with performing their tasks always in the correct manner to achieve the correct results always - right the first time.

RFT is the enabler to providing customers worldwide with accessible, high quality and advanced healthcare solutions which comply with cGMP requirements.

RFT in Practice
- Achieve excellence *rather than* "that's good enough"
- Prevent defects *rather than* "detect defects"
- Right first time *rather than* rework

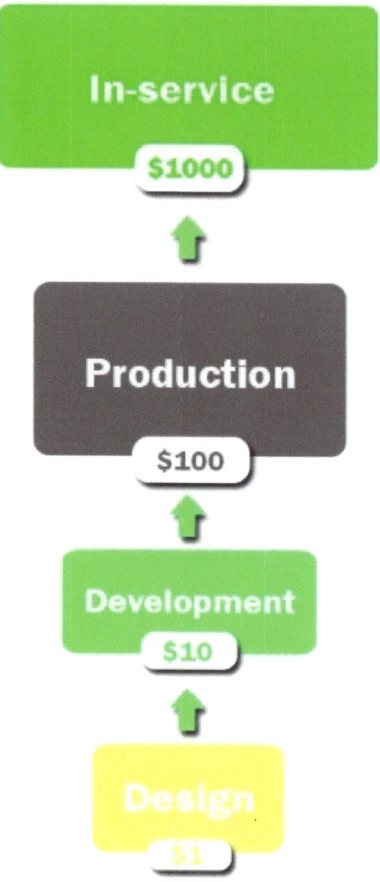

Figure: The increasing cost of correcting defects through the life-cycle of a product.

PDCA (Plan-Do-Check-Act)

PDCA (Plan–Do–Check–Act) is a four-step management tool often used in GLP and GMP environments

It sets in motion a repeatable and structured process-driven approach to solving problems and helps to drive consistent practices.

Plan

The Plan step is used to establish the objectives and desired goals of the proposed changes or modifications. Documenting these goals is important as it will drive the aspects of the next steps in the PDCA Process.

Do

Implement the plan and the changes identified. The "Do" step may require data collection and/or analysis prior to the implementation of changes. Training may also be required. Responsibilities should be clearly defined.

Check

Review results and analysis against the planned and expected results or goals.

Act

The act step ensures if any further corrective actions or modifications are noticed in the check step. The processes will require the person to "act" on the findings. However, any proposed changes are better captured by returning to the first step and restarting the process, either way, the application of PDCA will drive continuous improvement.

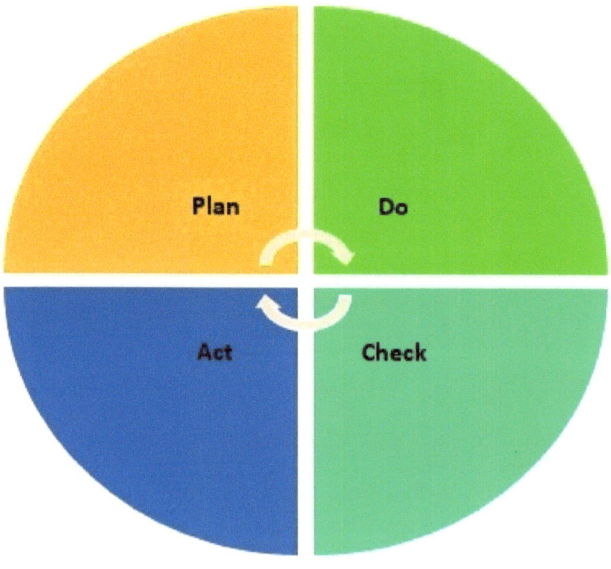

Figure: A Process Approach- PDCA Methodology

5s

5S is a Japanese methodology of organising and storing items in a work or lab environment. It has been adopted by many Western companies as a tool to help maintain standards and reduce errors and mix-ups. The "5s" represents each stage of the method.

Sort

Sorting out any items that are not in use and removing to a more appropriate area or to storage or the bin.

Set-in-Order

The idea of "Set-in-Order" is to be always organised. "A place for everything and everything in its place. "If we "set-in-order" we can help to make live processing and testing more efficient and reduce the risk of errors, omissions and accidents.

Shine

Regular cleaning is an important practice and it is always helpful to "Clean as you go."

Standardise

Implement standard practices through SOPs and training. Standardisation can also be applied to work station layout.

Sustain

Make it a habit! After implementing a 5s methodology, it is only effective if continuous efforts are made to "sustain" the changes.

Sort- Set-in-Order- Shine – Standardise – Sustain

www.ingramcontent.com/pod-product-compliance
Lightning Source LLC
Chambersburg PA
CBHW040854180526
45159CB00001B/418